Modern Sailboat Electrical System

A Comprehensive Guide to its Design, Construction, Boat Wiring and Outfitting

John K. Calder

Table of Contents

CHAPTER ONE 8

Introduction to Sailboat Electrical Systems 8

1.1 The Importance of Electrical Systems on Modern Sailboats..8

Overview of Sailboat Electrical Needs 8

1.1.1 Navigation and Communication 9

1.1.2 Lighting 10

1.1.3 Auxiliary Systems 10

1.1.4 Entertainment and Convenience 11

1.1.5 Safety Systems 11

Evolution of Electrical Systems in Sailboats.............12

1.1.6 Early Electrical Systems...................................12

1.1.7 Integration of Modern Technologies 12

1.1.8 Renewable Energy Sources 13

1.1.9 Smart Systems and Automation 13

1.2 Basic Electrical Concepts..14

Voltage, Current, Resistance, and Power 14

AC vs. DC Systems 16

1.2.5 Direct Current (DC) Systems 17

Advantages of DC systems:.......................................17

Disadvantages of DC systems: 18

1.2.6 Alternating Current (AC) Systems 18

Advantages of AC systems: 18

Disadvantages of AC systems: 19

1.3 Safety Considerations 19

Basic Safety Rules and Standards 20

1.3.4 Overcurrent Protection...................................21

Common Hazards and How to Avoid Them 22

1.3.10 Electromagnetic Interference (EMI) 24

CHAPTER TWO 25

Planning Your Electrical System................................25

Types of Voyages 25

Estimating Power Consumption 29

2.2 Budgeting and Cost Considerations 34

Cost Breakdown of Components 34

Balancing Cost and Quality 37

2.2.8 DIY vs. Professional Installation 37

2.3 System Design Principles 38

CHAPTER THREE 43

Power Sources 43

Types of Marine Batteries...43

3.1.3 Lithium-Ion Batteries 47

Battery Capacity and Bank Configuration 49

3.1.5 Calculating Total Capacity 50

Shore Power and Chargers 60

3.3 Renewable Energy Integration............................62

Solar and Wind Power System Design 63

Hybrid Systems 67

3.3.7 Energy Management..68

CHAPTER FOUR 70

Distribution Systems 70

4.1 Wiring and Cabling...70

Types of Marine-Grade Wire and Cable....................70

Proper Sizing and Selection 74

Fuses, Circuit Breakers, and Disconnects..................79

Designing for Accessibility and Expansion 88

CHAPTER FIVE 93

Electrical Components and Devices..........................93

5.1 Navigation and Communication Systems 93

GPS, Radar, and AIS 94

Installation: 94

5.2 Lighting Systems 100

Interior and Exterior Lighting..................................100

Water Makers, Refrigeration, and HVAC 106

Entertainment Systems and Other Conveniences ..112

CHAPTER SIX 116

Installation and Construction..................................116

- 6.1 Tools and Materials 116
- 6.3 Mounting and Securing Components 128
- 6.3.5 Protecting Against Moisture 132

CHAPTER SEVEN 134
- System Integration and Testing 134
- 7.1 Integrating Multiple Systems 134
- Ensuring Compatibility Between Systems 135
- 7.1.2 Interfacing Different Systems 136
- Communication Networks (NMEA 2000, Ethernet) 137

CHAPTER EIGHT 153
- Case Studies and Advanced Topics 153
- 8.1 Real-World Examples 153
- Case Study 1: Coastal Cruiser Retrofit 154
- 8.1.1 Project Overview 154
- 8.1.2 Planning and Design 154
- 8.1.3 Installation Process 155
- 8.1.4 Challenges and Solutions 156
- Case Study 2: Offshore Racing Yacht 158
- 8.1.6 Project Overview 158
- 8.1.7 Planning and Design 158
- 8.1.8 Installation Process 159
- 8.1.9 Challenges and Solutions 160
- 8.1.10 Outcome and Lessons Learned 160

8.2 Advanced Systems and Innovations 161

Smart Systems and Automation 162

8.2.1 Introduction to Smart Systems 162

Future Trends in Sailboat Electrical Systems 165

8.2.5 Energy Storage Innovations 166

8.2.6 Advanced Communication Systems..............167

Glossary of Terms 169

8.3.1 Key Terms and Definitions 169

Further Reading and Reference Materials..............171

8.3.3 Recommended Books and Articles 171

8.3.4 Online Resources 172

Websites: 172

Marine Electronics Journal:
www.marineelectronicsjournal.com 172

Summary...173

CHAPTER ONE

Introduction to Sailboat Electrical Systems

1.1 The Importance of Electrical Systems on Modern Sailboats

Overview of Sailboat Electrical Needs

Modern sailboats, whether used for leisurely day sailing, extended coastal cruising, or challenging offshore passages, rely heavily on sophisticated electrical systems to ensure functionality, safety, and comfort. These electrical systems power a wide range of equipment and devices, from essential navigation and communication tools to conveniences like refrigeration and entertainment systems. Understanding the electrical needs of a sailboat is the first step in designing an effective and reliable system.

1.1.1 Navigation and Communication

Navigation systems, such as GPS, radar, and autopilots, are crucial for safe and efficient sailing. GPS provides accurate positioning information, while radar helps detect other vessels and obstacles, particularly in low visibility conditions. Autopilots reduce the need for constant manual steering, allowing sailors to focus on other tasks.

Communication devices, including VHF radios, SSB (single sideband) radios, and satellite communication systems, are essential for maintaining contact with other vessels and shore-based facilities. These systems are vital for safety, weather updates, and general communication.

1.1.2 Lighting

Lighting is another critical aspect of a sailboat's electrical system. Navigation lights ensure that the vessel is visible to others, complying with international maritime regulations. Interior lighting enhances comfort and safety, allowing for activities below deck during nighttime or poor weather conditions.

1.1.3 Auxiliary Systems

Auxiliary systems like refrigeration, water makers, and HVAC (heating, ventilation, and air conditioning) significantly improve the quality of life aboard. Refrigeration keeps food fresh over long voyages, while water makers convert seawater into potable water, extending the duration a sailboat can remain at sea. HVAC systems maintain a comfortable environment inside the cabin regardless of external weather conditions.

1.1.4 Entertainment and Convenience

Modern sailors also expect entertainment systems such as audio players, televisions, and internet connectivity. These systems, while not essential for sailing, enhance the onboard experience, making long journeys more enjoyable.

1.1.5 Safety Systems

Safety systems, including bilge pumps, alarms, and electronic monitoring devices, are integral to a sailboat's electrical setup. Bilge pumps help remove water that enters the hull, preventing potential sinking. Alarms can warn of various issues like high water levels, smoke, or gas leaks, ensuring prompt action can be taken to avert disasters.

Evolution of Electrical Systems in Sailboats

The electrical systems on sailboats have evolved significantly over the past few decades. In the early days of sailing, boats had minimal or no electrical systems. Navigation was done using celestial bodies, and communication was limited to visual signals and sound.

1.1.6 Early Electrical Systems

The introduction of electrical systems began with basic lighting and simple navigation aids like compasses illuminated by electric lamps. As technology advanced, so did the complexity of onboard electrical systems.

1.1.7 Integration of Modern Technologies

The advent of GPS revolutionized navigation, providing accurate position data that could be integrated with digital charts and autopilots.

This allowed for precise route planning and hands-free steering. Similarly, the development of VHF radios enhanced communication capabilities, providing a reliable means to contact other vessels and shore stations.

1.1.8 Renewable Energy Sources

In recent years, there has been a significant push towards renewable energy sources on sailboats. Solar panels and wind turbines have become common, providing clean, sustainable power that reduces reliance on fossil fuels. These renewable sources help keep batteries charged, powering essential systems even when the engine is not running.

1.1.9 Smart Systems and Automation

The latest trend in sailboat electrical systems is the integration of smart technologies and automation. Systems can now be monitored and controlled remotely via smartphones or tablets.

Automated systems can manage power distribution, optimize energy usage, and provide real-time data on the status of various components.

1.2 Basic Electrical Concepts

To effectively design and manage a sailboat's electrical system, it is essential to understand some basic electrical concepts. This knowledge forms the foundation for troubleshooting problems and making informed decisions about component selection and system design.

Voltage, Current, Resistance, and Power

1.2.1 Voltage (V)

Voltage, measured in volts (V), is the electrical potential difference between two points. It is the driving force that pushes electric charges through a conductor.

On sailboats, systems typically operate at either 12V or 24V DC (direct current), although AC (alternating current) systems at 120V or 230V are also used, particularly when connected to shore power.

1.2.2 Current (I)

Current, measured in amperes (A), is the flow of electric charge through a conductor. It represents the rate at which electricity is flowing and is driven by the voltage across the conductor. High current can cause wires to heat up, so it's essential to use appropriately sized conductors to prevent overheating and potential fires.

1.2.3 Resistance (R)

Resistance, measured in ohms (Ω), is the opposition to the flow of electric current. It is determined by the material, length, and cross-sectional area of the conductor.

High resistance in a circuit can reduce the efficiency of the system and generate unwanted heat. In marine environments, corrosion can increase resistance, so using marine-grade materials is crucial.

1.2.4 Power (P)

Power, measured in watts (W), is the rate at which electrical energy is consumed or generated. It is calculated using the formula $P = V \times I$. Understanding power requirements is vital for sizing batteries, selecting charging systems, and ensuring that the electrical system can handle the load of all connected devices.

AC vs. DC Systems

Sailboat electrical systems typically use both DC and AC power, each with its own set of advantages and applications.

1.2.5 Direct Current (DC) Systems

DC systems are the backbone of most sailboat electrical setups. They are used for lighting, navigation equipment, communication devices, and other essential systems. DC power is supplied by batteries, which are charged through various means such as engine alternators, solar panels, and wind generators.

Advantages of DC systems:

- Simplicity: DC systems are relatively simple and easy to install and maintain.

- Safety: DC operates at lower voltages, reducing the risk of electric shock.

- Compatibility: Many marine devices are designed to run on DC power.

Disadvantages of DC systems:

- Voltage Drop: Over long distances, voltage drop can be significant, requiring larger conductors.

- Limited Power: DC systems are less efficient for high-power applications.

1.2.6 Alternating Current (AC) Systems

AC systems are used for high-power devices such as microwave ovens, air conditioning units, and shore power connections. AC power can be generated onboard using inverters, which convert DC power from batteries into AC power, or through generators that directly produce AC power.

Advantages of AC systems:

- Efficiency: AC is more efficient for transmitting power over long distances and for high-power applications.

- Versatility: Many household appliances and tools operate on AC power.

- Compatibility: Shore power is typically provided as AC, allowing for seamless integration when docked.

Disadvantages of AC systems:

- Complexity: AC systems are more complex and require careful design and installation.

- Safety: Higher voltages increase the risk of electric shock and require proper insulation and grounding.

1.3 Safety Considerations

Safety is paramount when working with electrical systems on sailboats. The marine environment presents unique challenges, such as exposure to moisture, salt, and vibration, which can lead to corrosion and component failure.

Adhering to safety standards and best practices is essential to ensure the longevity and reliability of the electrical system.

Basic Safety Rules and Standards
1.3.1 Use Marine-Grade Components

Always use components specifically designed for marine environments. Marine-grade wires, connectors, and devices are built to withstand harsh conditions and reduce the risk of corrosion and failure.

1.3.2 Follow Established Standards

Adhere to established standards such as those set by the American Boat and Yacht Council (ABYC) or the International Electrotechnical Commission (IEC). These standards provide guidelines for the safe design, installation, and maintenance of marine electrical systems.

1.3.3 Proper Grounding

Ensure that the electrical system is properly grounded. A good grounding system protects against electrical shock and helps prevent electrolysis, which can cause corrosion of metal parts in contact with water.

1.3.4 Overcurrent Protection

Install appropriate overcurrent protection devices, such as fuses and circuit breakers, to prevent electrical overloads and short circuits. These devices protect the wiring and components from damage due to excessive current flow.

1.3.5 Regular Maintenance

Perform regular maintenance and inspections of the electrical system. Look for signs of wear, corrosion, or damage, and address issues promptly to prevent failures.

Common Hazards and How to Avoid Them

1.3.6 Corrosion

Corrosion is a significant issue in marine environments due to the presence of salt and moisture. Use tinned copper wire and sealed connectors to minimize corrosion. Apply dielectric grease to connections to protect against moisture.

1.3.7 Overheating

Overheating can occur if wires are undersized or if connections are loose. Ensure that all wiring is properly sized for the current load and that connections are secure. Use thermal imaging tools to detect hotspots that might indicate a problem.

1.3.8 Electrical Shock

Electrical shock can result from improper insulation, grounding faults, or damaged components. Always turn off power before working on the electrical system. Use insulated tools and wear appropriate protective gear. Test for proper grounding and insulation regularly.

1.3.9 Fire Hazards

Electrical fires can be caused by short circuits, overloaded wires, or faulty components. Install smoke detectors and fire extinguishers in strategic locations. Use circuit breakers and fuses to prevent overloads. Regularly inspect wiring and replace any damaged or worn components.

1.3.10 Electromagnetic Interference (EMI)

EMI can disrupt the operation of navigation and communication equipment. Keep high-power AC lines and DC lines separate and well-shielded. Use twisted pair wires and ferrite cores to reduce interference.

By understanding the importance of electrical systems on modern sailboats, grasping basic electrical concepts, and adhering to safety standards, sailors can ensure that their vessels are equipped with reliable and efficient electrical systems. This knowledge not only enhances the functionality and comfort of the sailboat but also significantly contributes to the safety and well-being of everyone on board.

CHAPTER TWO

Planning Your Electrical System

2.1 Assessing Your Needs

Properly planning an electrical system for a sailboat requires a thorough understanding of the vessel's specific needs. This involves assessing the types of voyages you plan to undertake, estimating power consumption, and considering the unique requirements of your sailing lifestyle.

Types of Voyages

Different types of voyages place different demands on a sailboat's electrical system. Understanding these demands is crucial for designing a system that can meet your needs.

2.1.1 Day Sailing

Day sailing involves short trips, typically lasting a few hours to a single day, close to shore. The electrical demands for day sailing are relatively low since there is minimal need for extensive onboard systems.

- **Navigation Equipment:** Basic GPS, chartplotter, and possibly a VHF radio.

- **Lighting:** Navigation lights for safety and compliance with regulations.

- **Comfort:** Minimal need for auxiliary systems like refrigeration or extensive lighting.

For day sailors, the electrical system can be simple and focused on basic safety and navigation.

2.1.2 Coastal Cruising

Coastal cruising involves longer trips along the coastline, usually lasting a few days to a week. This type of voyage requires more robust electrical systems to support additional onboard activities and longer periods away from shore power.

- **Navigation and Communication:** Advanced GPS, radar, AIS, and VHF/SSB radios.

- **Lighting:** Comprehensive lighting for both navigation and interior comfort.

- **Auxiliary Systems:** Refrigeration, basic HVAC, and possibly a water maker.

- **Entertainment and Convenience:** Audio systems, basic internet connectivity, and cooking appliances.

Coastal cruisers need a well-rounded electrical system that balances capacity with efficiency.

2.1.3 Offshore Voyaging

Offshore voyaging involves extended periods at sea, often far from shore, requiring a highly reliable and self-sufficient electrical system. The demands for such voyages are the highest among the different types of sailing.

- **Navigation and Communication:** Full suite of navigation aids, including GPS, radar, AIS, depth sounders, and multiple communication systems (VHF, SSB, satellite).

- **Lighting:** Extensive lighting systems, including backup navigation lights.

- **Auxiliary Systems:** High-capacity refrigeration, advanced HVAC systems, water makers, and possibly washing machines.

- **Entertainment and Convenience:** Comprehensive entertainment systems, robust internet connectivity, and full galley appliances.

For offshore voyagers, redundancy and reliability are key, necessitating a robust electrical system capable of supporting long-term, self-sufficient living at sea.

Estimating Power Consumption

Accurately estimating power consumption is critical for designing an effective sailboat electrical system.

This involves listing all electrical devices, understanding their power requirements, and calculating the total energy consumption.

2.1.4 Inventory of Electrical Devices

Create a comprehensive list of all electrical devices on board. This should include navigation equipment, communication devices, lighting, auxiliary systems, entertainment gadgets, and any other electrical appliances.

Example of an inventory list:

- Navigation: GPS, radar, AIS, autopilot

- Communication: VHF radio, SSB radio, satellite phone

- Lighting: Navigation lights, interior lights, deck lights

- Auxiliary Systems: Refrigeration, HVAC, water maker

- Entertainment: Audio system, television, internet router

- Other: Bilge pumps, electric winches, galley appliances

2.1.5 Power Requirements of Each Device

Determine the power requirements for each device. This information is usually found in the device's specifications, often listed in watts (W) or amperes (A). For devices listed in watts, you can calculate the current draw using the formula:

$$I(A) = \frac{P(W)}{V(V)}$$

where I is the current in amperes, P is the power in watts, and V is the voltage.

2.1.6 Daily Usage Estimates

Estimate the daily usage for each device in hours. This will vary depending on the type of voyage and personal habits.

2.1.7 Calculating Daily Power Consumption

Multiply the power requirement of each device by the estimated daily usage to find the daily power consumption. Sum the consumption of all devices to get the total daily power consumption.

Example calculation:

- GPS: 5W, used 24 hours/day
    - Daily consumption: $5W \times 24h = 120Wh$
- Radar: 20W, used 4 hours/day
    - Daily consumption: $20W \times 4h = 80Wh$
- Refrigerator: 50W, used 12 hours/day
    - Daily consumption: $50W \times 12h = 600Wh$

Total daily consumption:

$$120Wh + 80Wh + 600Wh = 800Wh$$

2.2 Budgeting and Cost Considerations

Designing and installing a sailboat electrical system can be a significant investment. It is essential to balance cost with quality to ensure a reliable and efficient system without overspending.

Cost Breakdown of Components

Understanding the cost of various components helps in creating a realistic budget.

2.2.1 Batteries

Batteries are a significant part of the budget. Costs vary depending on the type (lead-acid, AGM, lithium-ion) and capacity.

- Lead-acid: $100-$200 per battery
- AGM: $200-$400 per battery
- Lithium-ion: $500-$1000 per battery

2.2.2 Charging Systems

Charging systems include alternators, solar panels, wind generators, and shore power chargers.

- Alternators: $300-$1000
- Solar panels: $100-$300 per panel
- Wind generators: $500-$1500
- Shore power chargers: $200-$500

2.2.3 Distribution Systems

Wiring, circuit breakers, fuses, and electrical panels are part of the distribution system.

- Marine-grade wire: $0.50-$2.00 per foot
- Circuit breakers and fuses: $10-$50 each
- Electrical panels: $100-$500

2.2.4 Electrical Components and Devices

Costs for navigation, communication, lighting, and auxiliary devices vary widely.

- GPS: $200-$1000
- Radar: $1000-$3000
- VHF radio: $100-$500
- LED lights: $10-$50 each
- Refrigeration: $500-$1500

2.2.5 Installation and Labor

If hiring a professional, labor costs can add significantly to the budget.

- Professional installation: $50-$100 per hour

Balancing Cost and Quality

2.2.6 Prioritize Critical Systems

Focus on the reliability and quality of critical systems such as navigation, communication, and safety devices. Invest in high-quality components for these systems to ensure reliability.

2.2.7 Evaluate Non-Essential Systems

For non-essential systems, such as entertainment and convenience devices, consider cost-effective options. Look for deals and consider second-hand equipment if it meets safety and performance standards.

2.2.8 DIY vs. Professional Installation

Consider your skills and experience when deciding between DIY installation and hiring a professional.

DIY can save money but may result in a less reliable system if not done correctly. Professional installation ensures quality and safety but increases costs.

2.3 System Design Principles

Designing an effective sailboat electrical system requires careful planning and adherence to key principles such as creating a detailed system diagram, ensuring redundancy, and maintaining reliability.

Creating a System Diagram

A system diagram provides a visual representation of the entire electrical system, helping to identify connections, components, and potential issues.

2.3.1 List All Components

Start by listing all components of the electrical system, including power sources, distribution elements, and end-use devices.

2.3.2 Draw the Diagram

Use a large sheet of paper or a digital drawing tool to create the diagram. Begin with the power sources at the top and work downwards, showing how power flows through the system to the various devices.

2.3.3 Indicate Connections

Clearly indicate all connections between components. Use standardized symbols for batteries, switches, fuses, circuit breakers, and other elements.

2.3.4 Label Components

Label each component with its name and specifications, such as voltage, current rating, and capacity. This helps in understanding the requirements and ensuring compatibility.

2.3.5 Review and Revise

Review the diagram for accuracy and completeness. Revise as necessary to ensure that all components and connections are correctly represented.

Redundancy and Reliability

Redundancy and reliability are crucial for ensuring that the electrical system remains functional even if a component fails.

2.3.6 Dual Power Sources

Consider having dual power sources, such as a combination of batteries and renewable energy sources (solar, wind). This provides backup power if one source fails.

2.3.7 Backup Systems

Implement backup systems for critical components. For example, have a spare VHF radio and handheld GPS in case the primary devices fail.

2.3.8 Parallel Circuits

Use parallel circuits for critical systems. This ensures that if one circuit fails, the others can continue to operate.

2.3.9 Quality Components

Invest in high-quality components, especially for critical systems. Cheap components are more likely to fail and can compromise the entire system.

2.3.10 Regular Testing

Regularly test all components and systems to identify potential issues before they lead to failures. Perform maintenance checks according to a schedule and address any problems immediately.

Planning an electrical system for a sailboat involves assessing the specific needs based on the type of voyages, accurately estimating power consumption, budgeting for components, and adhering to key design principles. By understanding these factors and following a structured approach, you can design a reliable and efficient electrical system that enhances the functionality, safety, and comfort of your sailboat. This comprehensive planning ensures that your vessel is well-equipped to handle the demands of modern sailing, whether for day trips, coastal cruising, or extended offshore voyages.

CHAPTER THREE

Power Sources

3.1 Batteries

Batteries are the heart of a sailboat's electrical system, storing energy for use when power generation is not possible. Understanding the different types of marine batteries, their capacity, and how to configure a battery bank is crucial for ensuring a reliable and efficient power source.

Types of Marine Batteries

Marine batteries come in various types, each with its own set of advantages and disadvantages. The most common types are lead-acid, AGM (Absorbent Glass Mat), and lithium-ion.

3.1.1 Lead-Acid Batteries

Lead-acid batteries are the traditional choice for marine applications. They are widely used due to their reliability and relatively low cost.

- **Flooded Lead-Acid Batteries:** These are the most common type and require regular maintenance, including checking the electrolyte levels and topping up with distilled water. They are durable and can withstand deep discharges, making them suitable for deep-cycle applications. However, they are heavier and bulkier compared to other types.

- **Sealed Lead-Acid Batteries:** These are maintenance-free and come in two main types: Gel and AGM. Gel batteries use a gel-like electrolyte, while AGM batteries use a fiberglass mat to absorb the electrolyte.

Advantages:

- Cost-effective
- Reliable and proven technology
- Available in a wide range of sizes and capacities

Disadvantages:

- Heavy and bulky
- Require maintenance (flooded type)
- Limited cycle life compared to newer technologies

3.1.2 AGM (Absorbent Glass Mat) Batteries

AGM batteries are a type of sealed lead-acid battery that offers several advantages over traditional flooded batteries.

- **Construction:** AGM batteries have a fiberglass mat that absorbs the electrolyte, making them spill-proof and maintenance-free.

- **Performance:** They provide higher power output and can handle deeper discharges better than flooded batteries. They also have a longer cycle life and can recharge faster.

Advantages:

- Maintenance-free

- Spill-proof and safer to install in various orientations

- Better performance and longer life than flooded lead-acid batteries

Disadvantages:

- More expensive than flooded lead-acid batteries

- Sensitive to overcharging

3.1.3 Lithium-Ion Batteries

Lithium-ion batteries are becoming increasingly popular in marine applications due to their high energy density, long cycle life, and low weight.

- **Construction:** Lithium-ion batteries use lithium compounds as the electrolyte. They are much lighter and compact than lead-acid batteries, offering significant weight savings.

- **Performance:** They provide a very high cycle life, can be discharged deeper, and charge much faster. They also have a more consistent voltage output over their discharge cycle.

Advantages:

- High energy density and lightweight
- Long cycle life and deep discharge capability
- Fast charging and consistent voltage

Disadvantages:

- High initial cost
- Require sophisticated battery management systems (BMS) to ensure safety and longevity
- Sensitivity to extreme temperatures

Battery Capacity and Bank Configuration

The capacity and configuration of your battery bank are critical for meeting your sailboat's power needs. Proper sizing ensures that you have enough stored energy for your devices and systems without overloading the batteries.

3.1.4 Understanding Battery Capacity

Battery capacity is measured in ampere-hours (Ah) and represents the amount of energy a battery can store. For example, a 100Ah battery can theoretically provide 1 amp of current for 100 hours or 10 amps for 10 hours.

- **Depth of Discharge (DoD):** This is the percentage of the battery's capacity that has been used. Lead-acid batteries should not be discharged more than 50% to maximize their lifespan, while lithium-ion batteries can safely discharge up to 80-90%.

3.1.5 Calculating Total Capacity

To determine the total capacity needed for your sailboat, calculate the daily energy consumption (in amp-hours) and ensure that your battery bank can provide this without exceeding the recommended DoD.

Example calculation:

- Daily consumption: 200Ah
- Desired DoD for lead-acid batteries: 50%
- Required battery bank capacity: $\frac{200Ah}{0.5} = 400Ah$

For lithium-ion batteries with an 80% DoD:

- Required battery bank capacity: $\frac{200Ah}{0.8} = 250Ah$

3.1.6 Battery Bank Configuration

Batteries can be configured in series or parallel to achieve the desired voltage and capacity.

- **Series Configuration:** Connecting batteries in series increases the voltage while keeping the capacity the same.

For example, connecting two 12V 100Ah batteries in series will give you 24V 100Ah.

- **Parallel Configuration:** Connecting batteries in parallel increases the capacity while keeping the voltage the same. For example, connecting two 12V 100Ah batteries in parallel will give you 12V 200Ah.

- **Series-Parallel Configuration:** This combines both methods to achieve higher voltage and capacity. For instance, four 12V 100Ah batteries can be configured in series-parallel to create a 24V 200Ah system.

3.1.7 Balancing the Battery Bank

Ensure that all batteries in a bank are of the same type, age, and capacity to maintain balance and efficiency.

Regularly check and balance the bank to prevent any single battery from being overcharged or deeply discharged.

3.2 Charging Systems

Charging systems are essential for replenishing the energy in your batteries. Multiple charging sources ensure that your batteries remain charged regardless of the conditions.

Alternators and Engine-Driven Generators

Alternators and engine-driven generators are primary charging sources, especially when the sailboat's engine is running.

3.2.1 Alternators

Alternators are installed on the engine and generate electricity while the engine is running. They are efficient and reliable but depend on the engine running, which may not always be practical.

- **Types of Alternators:** Standard alternators and high-output alternators. High-output alternators are designed to provide more charging power and are suitable for larger battery banks.

- **Regulators:** Alternators require voltage regulators to ensure that the batteries are charged correctly without overcharging. Modern multi-stage regulators optimize the charging process by adjusting the voltage and current based on the battery's state of charge.

Advantages:

- Reliable and efficient charging while the engine is running
- High-output alternators can quickly charge large battery banks

Disadvantages:

- Dependent on engine runtime
- Fuel consumption and engine wear

3.2.2 Engine-Driven Generators

Engine-driven generators are standalone units that produce AC power, which can be converted to DC power using a battery charger. They are suitable for larger vessels with significant power needs.

- **Types of Generators:** Diesel and gasoline generators. Diesel generators are more efficient and durable, making them the preferred choice for marine applications.

- **Output:** Generators can provide a substantial amount of power, suitable for charging large battery banks and running high-power devices.

Advantages:

- High power output for large battery banks and high-demand devices
- Can operate independently of the main engine

Disadvantages:

- High initial cost and maintenance
- Fuel consumption and noise

Solar Panels and Wind Generators

Renewable energy sources such as solar panels and wind generators are becoming increasingly popular for their ability to provide sustainable power without fuel costs.

3.2.3 Solar Panels

Solar panels convert sunlight into electrical energy, making them an excellent source of renewable power for sailboats.

- **Types of Solar Panels:** Monocrystalline, polycrystalline, and thin-film. Monocrystalline panels are the most efficient and compact, making them suitable for limited deck space.

- **Installation:** Panels can be mounted on the deck, bimini, or dedicated frames. Ensure they are installed in a location with maximum sun exposure and minimal shading.

- **Charge Controllers:** Solar panels require charge controllers to regulate the voltage and current going into the batteries, preventing overcharging. MPPT (Maximum Power Point Tracking) controllers are more efficient than PWM (Pulse Width Modulation) controllers.

Advantages:

- Sustainable and environmentally friendly
- Silent operation with no fuel costs
- Low maintenance

Disadvantages:

- Dependent on sunlight availability
- Initial installation cost

3.2.4 Wind Generators

Wind generators convert wind energy into electrical power, providing a consistent source of renewable energy, especially in windy conditions.

- **Types of Wind Generators:** Horizontal-axis and vertical-axis wind turbines. Horizontal-axis turbines are more common and efficient for marine applications.

- **Installation:** Mount wind generators on a sturdy mast or pole, ensuring they are above deck obstructions to maximize wind exposure.

- **Regulators:** Wind generators also require charge controllers to manage the charging process and prevent overcharging.

Advantages:

- Provides power day and night, as long as there is wind

- Complements solar panels, especially in cloudy or winter conditions

Disadvantages:

- Dependent on wind availability
- Noise and vibration
- Initial installation cost

Shore Power and Chargers

Shore power allows you to connect to the electrical grid when docked, providing a reliable source of AC power for charging batteries and running onboard systems.

3.2.5 Shore Power Connections

Shore power connections provide a direct link to the marina's electrical grid. They typically provide 120V or 240V AC power.

- **Connection Types:** Standard marine shore power connections are 30A or 50A. Ensure you have the appropriate cables and connectors for your vessel.

- **Safety:** Use a shore power inlet with proper grounding and circuit protection to prevent electrical hazards.

3.2.6 Battery Chargers

Battery chargers convert AC power from shore power or generators to DC power for charging batteries.

- **Types of Chargers:** Single-stage, multi-stage, and smart chargers. Multi-stage and smart chargers optimize the charging process, extending battery life.

- **Capacity:** Choose a charger with sufficient output to match your battery bank's size. Chargers are rated in amps, and a higher amp rating means faster charging.

Advantages:

- Reliable and consistent power source
- Can charge batteries and run onboard systems simultaneously

Disadvantages:

- Dependent on marina availability
- Requires proper connectors and safety precautions

3.3 Renewable Energy Integration

Integrating renewable energy sources into your sailboat's electrical system provides sustainable and reliable power, reducing dependence on fossil fuels and shore power.

Solar and Wind Power System Design

Designing a solar and wind power system involves selecting the right components and integrating them into your existing electrical system.

3.3.1 Assessing Energy Needs

Start by calculating your daily energy consumption, as discussed in Chapter 2. This will help determine the capacity needed for your renewable energy sources.

3.3.2 Selecting Solar Panels

Choose solar panels based on available space, energy needs, and budget.

- **Panel Capacity:** The total capacity (in watts) should be sufficient to meet your daily energy consumption.

For example, if your daily consumption is 500Wh, a 100W solar panel producing 5 hours of peak sunlight will generate $100W \times 5h = 500Wh$.

- **Number of Panels:** Determine the number of panels required based on their individual capacities. For instance, if using 100W panels, you would need five panels to generate 500Wh per day.

3.3.3 Selecting Wind Generators

Choose a wind generator based on average wind conditions and energy needs.

- **Generator Capacity:** The capacity (in watts) should complement the solar panels. If solar panels provide 60% of your energy needs, the wind generator should cover the remaining 40%.

- **Mounting Location:** Install the wind generator in a location with minimal obstructions and maximum wind exposure.

3.3.4 Charge Controllers

Use MPPT charge controllers for both solar panels and wind generators to optimize energy conversion and battery charging.

- **Solar Charge Controllers:** Choose an MPPT controller with a capacity matching your solar panel array. For instance, a 500W array with a 12V system would need a controller rated for at least 40A $\left(\frac{500W}{12V} = 41.67A\right)$.

-

- **Wind Charge Controllers:** Select a controller that matches the wind generator's output and integrates with your existing system.

3.3.5 System Integration

Integrate solar panels and wind generators into your existing electrical system.

- **Combiner Box:** Use a combiner box to merge multiple solar panel connections into a single output.

- **Battery Bank Connection:** Connect the charge controllers to the battery bank, ensuring proper voltage and current settings.

- **Monitoring:** Install a monitoring system to track the performance and health of your renewable energy sources and battery bank.

Hybrid Systems

Hybrid systems combine multiple power sources to ensure a consistent and reliable energy supply.

3.3.6 Combining Renewable and Engine-Driven Power

Integrate renewable energy sources with alternators and generators to create a hybrid system.

- **Hybrid Controller:** Use a hybrid charge controller to manage inputs from both renewable and engine-driven sources.

- **Priority Settings:** Configure the system to prioritize renewable energy, using engine-driven power only when necessary.

3.3.7 Energy Management

Implement an energy management system to optimize power generation and consumption.

- **Smart Inverters:** Use smart inverters to convert DC to AC power efficiently, integrating with renewable and engine-driven sources.

- **Load Management:** Prioritize essential loads and manage non-essential loads to ensure efficient energy use.

3.3.8 Backup Power

Ensure redundancy with backup power sources.

- **Battery Bank:** Maintain a sufficient battery bank capacity to handle peak loads and emergencies.

- **Backup Generator:** Consider a portable backup generator for emergencies or extended periods of low renewable energy production.

Power sources are the foundation of a sailboat's electrical system. By understanding the different types of batteries, charging systems, and integrating renewable energy sources, you can design a reliable and efficient power system. Proper planning and integration of multiple power sources ensure that your sailboat remains powered under various conditions, enhancing your sailing experience and reducing dependence on fossil fuels. This chapter provides a comprehensive guide to selecting, configuring, and integrating power sources, setting the stage for a robust and sustainable sailboat electrical system.

CHAPTER FOUR
Distribution Systems
4.1 Wiring and Cabling

The wiring and cabling system is the backbone of your sailboat's electrical system. It ensures that power is distributed efficiently and safely to all the devices and systems on board. Proper selection, sizing, and installation of marine-grade wire and cable are critical to avoid voltage drops, overheating, and potential electrical failures.

Types of Marine-Grade Wire and Cable

Marine-grade wire and cable are specifically designed to withstand the harsh marine environment, including exposure to moisture, salt, vibration, and temperature fluctuations.

The following types of wires and cables are commonly used in sailboat electrical systems:

4.1.1 Stranded Copper Wire

Stranded copper wire is the standard for marine applications due to its flexibility and ability to withstand vibration and movement.

- **Construction:** Composed of multiple small strands of copper wire twisted together. This construction allows the wire to flex without breaking.

- **Tinned Copper:** Tinned copper wire is coated with a thin layer of tin to resist corrosion. This is especially important in the marine environment where exposure to saltwater can lead to rapid corrosion of bare copper.

4.1.2 Battery Cables

Battery cables are thicker, more robust cables designed to handle the high currents associated with batteries and starter motors.

- **Construction:** Typically made of stranded copper, these cables are heavily insulated to prevent damage and ensure safety.
- **Sizes:** Common sizes range from 4 AWG (American Wire Gauge) to 4/0 AWG, depending on the current requirements of the system.

4.1.3 Coaxial Cable

Coaxial cable is used for transmitting radio frequency signals, such as those from VHF radios and antennas.

- **Construction:** Consists of a central conductor, an insulating layer, a metallic shield, and an outer insulating layer. This design minimizes signal loss and interference.

- **Types:** RG-58 and RG-213 are common types used in marine applications, with RG-213 offering lower signal loss over longer distances.

4.1.4 Twisted Pair Cable

Twisted pair cable is used for data and communication lines, such as NMEA 2000 networks and Ethernet connections.

- **Construction:** Consists of pairs of wires twisted together to reduce electromagnetic interference (EMI).

- **Shielded vs. Unshielded:** Shielded twisted pair (STP) cables provide additional protection against EMI, while unshielded twisted pair (UTP) cables are more flexible and easier to install.

Proper Sizing and Selection

Proper sizing of wire and cable is crucial to ensure safety, efficiency, and longevity of the electrical system. Undersized wires can lead to excessive voltage drops and overheating, while oversized wires can be unnecessarily expensive and difficult to install.

4.1.5 Determining Wire Size

Wire size is determined by the current (in amperes) that the wire needs to carry and the distance (in feet) the current must travel. The American Boat and Yacht Council (ABYC) provides guidelines for selecting wire size based on these factors.

- **Current (Amperage):** Calculate the total current draw of all devices on the circuit. For example, if a circuit supplies power to a 10A device and a 5A device, the total current is 15A.

- **Distance:** Measure the distance from the power source (e.g., battery) to the device and back. This is the total length of the circuit.

4.1.6 Voltage Drop

Voltage drop is the reduction in voltage that occurs as electrical current travels through a wire. Excessive voltage drop can lead to poor performance or damage to electrical devices.

- **Acceptable Voltage Drop:** For most 12V systems, a voltage drop of 3% is considered acceptable for critical circuits (e.g., navigation lights), while a 10% drop is acceptable for non-critical circuits (e.g., cabin lights).

4.1.7 Using a Wire Size Calculator

Wire size calculators are tools that help determine the appropriate wire size based on current, distance, and acceptable voltage drop. Input the values, and the calculator will provide the recommended wire size.

Example calculation:

- Current: 15A
- Distance: 20 feet (one way)
- Voltage: 12V
- Acceptable voltage drop: 3%

Using these values, the calculator might recommend a 10 AWG wire.

4.1.8 Selecting the Right Type of Wire

- **Primary Wiring:** Use stranded, tinned copper wire for most applications. Ensure the wire is rated for the appropriate voltage and has adequate insulation.

- **Battery Cables:** Use heavy-duty, flexible battery cables for connections to the battery bank. These cables should be oversized to handle the high currents associated with starting and charging.

- **Specialty Cables:** Use coaxial cable for RF signals and twisted pair cable for data communication.

4.2 Circuit Protection

Circuit protection devices are essential for safeguarding the electrical system from overcurrent, short circuits, and other electrical faults. Proper selection and installation of fuses, circuit breakers, and disconnects are critical for maintaining safety and reliability.

Fuses, Circuit Breakers, and Disconnects

4.2.1 Fuses

Fuses are sacrificial devices that protect circuits by breaking the connection when the current exceeds a specified limit. They are simple, reliable, and inexpensive.

- **Types of Fuses:** Blade fuses, glass tube fuses, and ANL fuses are commonly used in marine applications. Each type has different characteristics and applications.

- **Rating:** Select a fuse with a current rating slightly higher than the maximum expected current in the circuit. For example, if a circuit is expected to carry 10A, a 15A fuse would be appropriate.

- **Installation:** Install fuses as close to the power source as possible to protect the entire circuit. Use fuse holders or blocks for secure and accessible installation.

4.2.2 Circuit Breakers

Circuit breakers provide overcurrent protection similar to fuses but can be reset after tripping, making them more convenient for frequently accessed circuits.

- **Types of Circuit Breakers:** Thermal, magnetic, and thermal-magnetic circuit breakers are common in marine applications. Thermal breakers respond to heat generated by overcurrent, while magnetic breakers respond to magnetic fields created by current flow.

- **Rating:** Select a circuit breaker with a current rating slightly higher than the maximum expected current in the circuit.

Ensure the breaker can handle the voltage and interrupt current for the application.

- **Installation:** Install circuit breakers in accessible locations, such as electrical panels, for easy resetting. Use breaker panels with proper labeling for organization.

4.2.3 Disconnects

Disconnect switches allow for quickly disconnecting power to a circuit or the entire electrical system. They are essential for safety during maintenance and emergencies.

- **Types of Disconnects:** Battery disconnect switches, main disconnect switches, and emergency cut-off switches are commonly used.

- **Installation:** Install disconnect switches in easily accessible locations. For battery disconnects, place them near the battery bank to isolate the entire system quickly.

Overcurrent Protection Strategies

Overcurrent protection strategies ensure that all circuits are adequately protected against excessive current, preventing damage to wiring and devices.

4.2.4 Protecting Individual Circuits

Each individual circuit should have its own overcurrent protection device (fuse or circuit breaker) sized according to the current capacity of the wire and the expected load.

- **Branch Circuits:** Install fuses or circuit breakers on each branch circuit to protect the wires and connected devices.

- **Main Circuit:** Install a main circuit breaker or fuse to protect the main wiring from the battery to the distribution panel.

4.2.5 Sizing Overcurrent Protection

Properly size fuses and circuit breakers to match the wire size and load.

- **Wire Size:** Ensure the overcurrent protection device is rated lower than the current-carrying capacity of the wire. For example, a 14 AWG wire typically carries up to 15A, so use a fuse or breaker rated at or below 15A.

- **Load Requirements:** The overcurrent protection device should be rated slightly above the maximum expected current of the load. This allows for normal operation without nuisance trips while still providing protection.

4.2.6 Coordinating Protection Devices

Coordinate overcurrent protection devices to ensure that the correct device trips first in the event of a fault.

- **Selectivity:** Choose devices with appropriate time-current characteristics to ensure that branch circuit protection devices trip before the main protection device.

- **Cascade Protection:** Implement cascading protection by using smaller devices for individual circuits and larger devices for main circuits. This provides a layered approach to protection.

4.3 Electrical Panels and Switches

Electrical panels and switches organize and control the distribution of power throughout the sailboat. Proper design and installation ensure easy access, safety, and future expandability.

Main and Sub-Panels

Electrical panels serve as central points for circuit protection and power distribution. Main panels handle primary distribution, while sub-panels manage specific areas or systems.

4.3.1 Main Panels

The main electrical panel is the primary distribution point for the boat's electrical system.

- **Components:** Main breaker, individual circuit breakers or fuses, voltage and current meters, and indicator lights.

- **Location:** Install the main panel in a dry, accessible location, typically near the navigation station or helm.

- **Organization:** Label all breakers and circuits clearly for easy identification and troubleshooting.

4.3.2 Sub-Panels

Sub-panels distribute power to specific areas or systems, such as the galley, cabins, or navigation equipment.

- **Components:** Smaller versions of the main panel with circuit breakers or fuses for individual circuits.

- **Location:** Install sub-panels near the areas they serve to minimize wiring runs and provide local control.

- **Integration:** Connect sub-panels to the main panel with appropriately sized wire and overcurrent protection.

4.3.3 Panel Selection and Sizing

Choose electrical panels based on the number of circuits, current capacity, and expandability.

- **Number of Circuits:** Select a panel with enough circuit breakers or fuse slots to accommodate current and future needs. Plan for at least 20-30% more circuits than initially required for future expansion.

- **Current Capacity:** Ensure the panel can handle the total current load. For example, a panel serving a 100A system should have a main breaker rated for 100A or more.

- **Expandability:** Choose panels with modular designs that allow for easy addition of circuits and breakers.

Designing for Accessibility and Expansion

Designing the electrical system with accessibility and future expansion in mind ensures ease of maintenance and upgrades.

4.3.4 Accessibility

Ensure that all electrical panels, switches, and wiring are easily accessible for inspection, maintenance, and troubleshooting.

- **Location:** Install panels and switches in locations that are easy to reach and free from obstructions. Avoid areas that are difficult to access or prone to moisture and corrosion.

- **Labeling:** Clearly label all circuits, switches, and connections to facilitate identification and troubleshooting. Use durable, water-resistant labels or engraved plaques.

4.3.5 Future Expansion

Design the electrical system with future expansion in mind to accommodate additional equipment and upgrades.

- **Spare Circuits:** Include spare circuits in the panels for future additions. This reduces the need for major modifications when adding new devices.

- **Conduit and Cable Trays:** Install conduit or cable trays with extra capacity to accommodate additional wiring. This simplifies future wiring runs and maintains organization.

- **Modular Components:** Use modular components and connectors that allow for easy expansion and reconfiguration. This includes modular breaker panels, plug-and-play connectors, and quick-release mounts.

4.3.6 Redundancy and Reliability

Implement redundancy and reliability measures to ensure continuous operation and minimize downtime.

- **Dual Panels:** Consider installing dual main panels or critical sub-panels to provide redundancy. This ensures that a failure in one panel does not disable the entire system.

- **Backup Systems:** Integrate backup power sources, such as an additional battery bank or generator, to maintain power during primary system failures.

- **Robust Components:** Use high-quality, marine-grade components that are designed to withstand the harsh marine environment. This includes corrosion-resistant connectors, waterproof switches, and vibration-resistant mounts.

A well-designed distribution system is essential for the reliable and safe operation of a sailboat's electrical system. By selecting the appropriate marine-grade wire and cable, implementing proper circuit protection, and designing accessible and expandable electrical panels, you can ensure that power is efficiently distributed to all devices and systems on board. This chapter provides a detailed guide to wiring, circuit protection, and electrical panel design, laying the foundation for a robust and efficient sailboat electrical system.

Proper planning and execution in these areas will enhance the overall functionality and safety of your vessel, allowing you to enjoy your time on the water with confidence.

CHAPTER FIVE

Electrical Components and Devices

Modern sailboats are equipped with a variety of electrical components and devices that enhance safety, navigation, comfort, and overall experience on the water. This chapter delves into the various systems and devices, providing detailed step-by-step explanations on their functionality, installation, and integration into the sailboat's electrical system.

5.1 Navigation and Communication Systems

Navigational and communication systems are essential for the safety and efficiency of sailing. They provide critical information about the boat's position, surroundings, and allow for effective communication with other vessels and shore stations.

GPS, Radar, and AIS
5.1.1 GPS (Global Positioning System)

GPS is a satellite-based navigation system that provides accurate location and time information.

- **Components:** GPS receiver, antenna, and display unit (chartplotter or multifunction display).

- **Functionality:** The GPS receiver calculates the boat's position by triangulating signals from multiple satellites. This information is displayed on the chartplotter, providing real-time location, speed, and course.

Installation:
1. **Select Location:** Mount the GPS antenna in a location with a clear view of the sky, free from obstructions.

2. **Run Cables:** Route the antenna cable to the GPS receiver, avoiding sharp bends and potential chafe points.

3. **Connect Power:** Connect the GPS receiver to the power supply, typically through a fused circuit on the main electrical panel.

4. **Mount Display:** Install the display unit at the helm or navigation station for easy viewing and operation.

5. **Test System:** Power on the system and verify that it acquires satellite signals and displays accurate position data.

5.1.2 Radar

Radar systems detect objects and measure their distance by bouncing radio waves off them, essential for navigation in low visibility conditions.

- **Components:** Radar scanner (antenna), display unit (radar screen), and connecting cables.

- **Functionality:** The radar scanner emits radio waves that reflect off objects, returning signals that are processed and displayed on the radar screen, showing the distance and bearing of objects.

- **Installation:**

    1. **Select Mounting Location:** Mount the radar scanner on the mast or a dedicated radar arch, ensuring it is elevated and unobstructed.

    2. **Run Cables:** Route the power and data cables from the scanner to the display unit, securing them along the mast or radar arch.

    3. **Connect Power:** Connect the radar system to the power supply via a fused circuit.

    4. **Mount Display:** Install the radar display at the helm or navigation station.

5. **Test System:** Power on the radar system, verify it rotates correctly, and check the display for accurate target acquisition.

5.1.3 AIS (Automatic Identification System)

AIS transmits and receives information about vessels, including position, course, speed, and vessel details, enhancing situational awareness and collision avoidance.

- **Components:** AIS transceiver, GPS antenna, VHF antenna, and display unit or integration with existing chartplotter.

- **Functionality:** The AIS transceiver sends and receives vessel information, which is displayed on the chartplotter or a dedicated display unit.

- **Installation:**

1. **Select Mounting Location:** Install the AIS transceiver in a dry, accessible location near the navigation station.

2. **Run Antenna Cables:** Connect the GPS and VHF antennas to the transceiver, routing cables to avoid interference.

3. **Connect Power:** Connect the AIS transceiver to the power supply using a fused circuit.

4. **Integrate Display:** Connect the AIS to the chartplotter or a dedicated display unit.

5. **Test System:** Power on the AIS, verify it acquires GPS signals, and check the display for vessel information.

5.2 Lighting Systems

Lighting systems on sailboats serve functional, safety, and aesthetic purposes. Proper installation and compliance with regulations are crucial for ensuring visibility and safety.

Interior and Exterior Lighting

5.2.1 Interior Lighting

Interior lighting enhances visibility and comfort inside the cabin.

- **Types of Lights:** Overhead lights, reading lights, strip lights, and courtesy lights.
- **Installation:**
    1. **Plan Lighting Layout:** Determine the locations and types of lights needed for different areas (e.g., galley, salon, berths).

2. **Run Wiring:** Route wiring through the boat's interior, using conduit or wire looms to protect and organize cables.

3. **Install Lights:** Mount the light fixtures, ensuring secure attachment and proper alignment.

4. **Connect Power:** Connect each light to the electrical system, using switches or dimmers as needed.

5. **Test Lighting:** Power on the system and verify that all lights function correctly.

5.2.2 Exterior Lighting

Exterior lighting includes deck lights, spreader lights, and courtesy lights for safe movement on deck at night.

- **Types of Lights:** Deck lights, spreader lights, transom lights, and cockpit lights.
- **Installation:**
    1. **Plan Lighting Layout:** Identify locations for exterior lights to ensure adequate illumination.
    2. **Run Wiring:** Route wiring through the boat's exterior, protecting it from weather and mechanical damage.
    3. **Install Lights:** Mount exterior light fixtures, using waterproof connectors and sealing all penetrations.
    4. **Connect Power:** Connect exterior lights to the power supply through switches or control panels.

5. **Test Lighting:** Power on the system and verify the functionality of all exterior lights.

Navigation Lights and Their Regulations

Navigation lights are crucial for signaling the boat's position and movement to other vessels, especially at night or in low visibility conditions. Compliance with international regulations (e.g., COLREGs) is mandatory.

5.2.3 Navigation Lights

Navigation lights indicate the sailboat's status (under sail or power) and direction of movement.

- **Types of Navigation Lights:**
    - **Masthead Light:** White light visible over 225 degrees, used when under power.

- o **Sidelights:** Red (port) and green (starboard) lights visible over 112.5 degrees each.
- o **Stern Light:** White light visible over 135 degrees.
- o **Anchor Light:** White light visible over 360 degrees, used when at anchor.

- **Installation:**

1. **Plan Installation:** Identify locations for each navigation light, ensuring compliance with visibility and positioning regulations.

2. **Run Wiring:** Route wiring through the mast, deck, and hull, protecting it from weather and mechanical damage.

3. **Install Lights:** Mount the navigation light fixtures securely, using waterproof connectors and sealing all penetrations.

4. **Connect Power:** Connect navigation lights to the power supply through dedicated switches or an automated control panel.

5. **Test Lights:** Power on the system and verify that all navigation lights function correctly and meet visibility requirements.

5.2.4 Compliance with Regulations

Ensure that all navigation lights comply with international regulations, such as the International Regulations for Preventing Collisions at Sea (COLREGs).

- **Visibility Requirements:** Navigation lights must be visible at specified distances depending on the boat's length.

- **Positioning:** Lights must be positioned to indicate the boat's status and movement accurately.

- **Color and Intensity:** Use lights with the correct color and intensity to meet regulatory standards.

5.3 Auxiliary Devices

Auxiliary devices add comfort, convenience, and functionality to the sailboat, enhancing the overall sailing experience.

Water Makers, Refrigeration, and HVAC

5.3.1 Water Makers

Water makers (desalination units) convert seawater into potable water, providing an independent freshwater supply.

- **Components:** Pre-filter, high-pressure pump, reverse osmosis membrane, post-filter, and control unit.

- **Functionality:** Seawater is pressurized and passed through a reverse osmosis membrane, removing salts and impurities to produce freshwater.

- **Installation:**

    1. **Select Location:** Install the water maker in a dry, accessible location near the water intake and storage tanks.

    2. **Run Plumbing:** Connect the seawater intake, brine discharge, and freshwater output to the respective systems.

3. **Connect Power:** Connect the water maker to the power supply, typically through a dedicated circuit with a fuse or breaker.

4. **Mount Control Unit:** Install the control unit in an accessible location for monitoring and operation.

5. **Test System:** Power on the system, check for leaks, and verify that it produces potable water.

5.3.2 Refrigeration

Refrigeration systems keep food and beverages cold, essential for extended voyages.

- **Types of Refrigeration:** Compressor-based units, thermoelectric coolers, and absorption refrigerators.

- **Installation:**
    1. **Select Location:** Install the refrigerator in a ventilated area to ensure efficient heat dissipation.
    2. **Run Wiring:** Connect the refrigerator to the power supply, using appropriately sized wire and overcurrent protection.
    3. **Install Insulation:** Ensure the refrigerator is well-insulated to minimize energy consumption.
    4. **Connect Power:** Connect the refrigerator to the electrical system through a dedicated circuit.
    5. **Test System:** Power on the refrigerator and verify that it maintains the desired temperature.

5.3.3 HVAC (Heating, Ventilation, and Air Conditioning)

HVAC systems regulate the cabin temperature and provide ventilation, enhancing comfort in varying weather conditions.

- **Components:** Air handler, compressor, condenser, ducting, and control unit.

- **Functionality:** The system circulates air, cools or heats it, and distributes it through the cabin via ducting.

- **Installation:**

    1. **Select Location:** Install the HVAC components in accessible, ventilated locations to ensure efficient operation.

    2. **Run Ducting:** Route ducting through the boat to distribute conditioned air to various areas.

3. **Connect Plumbing:** Connect the HVAC system to the seawater cooling intake and discharge, if applicable.

4. **Connect Power:** Connect the HVAC system to the power supply, using appropriately sized wire and overcurrent protection.

5. **Mount Control Unit:** Install the control unit in an accessible location for temperature and system control.

6. **Test System:** Power on the HVAC system, verify proper operation, and check for any leaks or issues.

Entertainment Systems and Other Conveniences

5.3.4 Entertainment Systems

Entertainment systems provide audio and visual enjoyment, enhancing the onboard experience.

- **Components:** Stereo system, speakers, television, DVD/Blu-ray player, and antenna.

- **Installation:**

    1. **Select Components:** Choose marine-rated components designed to withstand the marine environment.

    2. **Plan Layout:** Determine locations for the stereo, speakers, and other components to ensure optimal sound and viewing.

3. **Run Wiring:** Route wiring for power, speakers, and antennas through the boat, protecting it from damage.

4. **Install Components:** Mount the stereo, speakers, and other components securely, using waterproof connectors where necessary.

5. **Connect Power:** Connect the entertainment system to the power supply, typically through a dedicated circuit.

6. **Test System:** Power on the system, adjust settings, and verify sound and picture quality.

5.3.5 Other Conveniences

Additional devices can enhance comfort and convenience on board, including:

- **Coffee Makers and Small Appliances:** Use dedicated outlets and circuits for high-power appliances.

- **Inverters and Chargers:** Install inverters to provide AC power from DC batteries, and chargers to maintain battery health.

- **Fans and Dehumidifiers:** Improve air circulation and reduce humidity for a more comfortable cabin environment.

This chapter has provided an in-depth exploration of the various electrical components and devices commonly found on modern sailboats. From essential navigation and communication systems to lighting, auxiliary devices, and entertainment systems, each component plays a crucial role in enhancing the safety, functionality, and comfort of the vessel. Proper selection, installation, and integration of these systems are critical to ensure reliable and efficient operation, allowing you to enjoy a

seamless and enjoyable sailing experience. By following the detailed step-by-step explanations provided, you can confidently outfit your sailboat with the necessary electrical components and devices, ensuring a well-equipped and comfortable vessel ready for any adventure on the water.

CHAPTER SIX

Installation and Construction

Proper installation and construction of your sailboat's electrical system is crucial for ensuring safety, reliability, and efficiency. This chapter provides detailed step-by-step explanations on the tools and materials required, wiring techniques, and methods for mounting and securing components. By following these guidelines, you can achieve a well-constructed and dependable electrical system.

6.1 Tools and Materials

Having the right tools and materials is essential for performing electrical work on a sailboat. This section covers the essential tools and how to select and source quality materials.

Essential Tools for Electrical Work

6.1.1 Basic Hand Tools

Basic hand tools are the foundation of any electrical work. These tools include:

- **Screwdrivers:** A set of flathead and Phillips screwdrivers of various sizes for securing terminals and connectors.

- **Pliers:** Needle-nose, side-cutting, and lineman's pliers for gripping, cutting, and twisting wires.

- **Wire Strippers:** Tools for stripping insulation from wires of different gauges without damaging the conductors.

- **Crimping Tools:** Specialized tools for crimping terminals and connectors securely onto wires.

- **Wrenches and Sockets:** A set of wrenches and sockets for fastening components and making secure connections.

- **Multimeter:** A digital multimeter for measuring voltage, current, resistance, and continuity to diagnose and troubleshoot electrical issues.

6.1.2 Specialized Electrical Tools

Specialized electrical tools make specific tasks easier and more precise:

- **Cable Cutters:** Heavy-duty cutters for cleanly cutting thicker cables.

- **Heat Gun:** A heat gun for shrinking heat-shrink tubing to insulate and protect wire connections.

- **Fish Tape:** A tool for pulling wire through conduit or tight spaces.

- **Label Maker:** A label maker for creating durable, easy-to-read labels for wires, panels, and components.

- **Cable Ties and Clamps:** Cable ties and clamps for organizing and securing wiring runs.

Selecting and Sourcing Materials

6.1.3 Wire and Cable

Choosing the right wire and cable is crucial for safety and performance:

- **Marine-Grade Wire:** Use marine-grade wire, which has tinned copper conductors for corrosion resistance and flexibility. Common sizes are 10 AWG for heavy loads, 12 AWG for medium loads, and 14-16 AWG for light loads.

- **Battery Cables:** Use large-diameter cables (2/0, 1/0, 2 AWG) for battery connections to handle high current loads with minimal voltage drop.

- **Shielded Cable:** Use shielded cable for sensitive electronics to prevent electromagnetic interference (EMI).

6.1.4 Connectors and Terminals

Quality connectors and terminals ensure secure and reliable connections:

- **Ring Terminals:** Use ring terminals for connecting wires to screw posts and terminals.

- **Butt Connectors:** Use butt connectors for splicing wires together securely.

- **Heat-Shrink Connectors:** Use heat-shrink connectors with built-in adhesive for watertight and corrosion-resistant connections.

6.1.5 Circuit Protection

Circuit protection devices are essential for preventing electrical fires and equipment damage:

- **Fuses:** Use marine-grade fuses with appropriate amperage ratings for each circuit.

- **Circuit Breakers:** Use circuit breakers for easily resettable overcurrent protection.

- **Fuse Holders and Panels:** Use fuse holders and panels designed for marine environments to organize and protect fuses.

6.1.6 Mounting Hardware

Quality mounting hardware ensures components are securely attached:

- **Stainless Steel Fasteners:** Use stainless steel screws, bolts, and nuts for corrosion resistance.

- **Mounting Brackets:** Use mounting brackets and plates designed for specific components to ensure secure attachment.

6.2 Wiring Techniques

Proper wiring techniques are essential for a safe and reliable electrical system. This section covers the best practices for running wire, making connections, and terminating wires.

Proper Techniques for Running Wire

6.2.1 Planning the Wiring Layout

Before running any wire, plan the wiring layout to ensure efficient routing and minimize interference:

- **Create a Wiring Diagram:** Develop a detailed wiring diagram showing all components, wire runs, and connections.

- **Group Wires by Function:** Group wires by function (e.g., power, lighting, navigation) to simplify troubleshooting and maintenance.

- **Avoid Interference:** Route power and signal wires separately to minimize EMI.

6.2.2 Routing Wires

Proper wire routing is essential for protecting wires and ensuring reliable connections:

- **Use Conduit and Cable Trays:** Run wires through conduit or cable trays to protect them from mechanical damage and organize the installation.

- **Avoid Sharp Bends:** Avoid sharp bends in wire runs, which can damage the insulation and conductors.

- **Secure Wires:** Use cable ties, clamps, and straps to secure wires at regular intervals, preventing movement and chafing.

- **Label Wires:** Clearly label all wires at both ends to facilitate identification and troubleshooting.

Connections and Terminations

6.2.3 Making Connections

Secure and reliable connections are critical for preventing electrical failures:

- **Strip the Wire:** Use a wire stripper to remove the appropriate length of insulation without damaging the conductors.

- **Select the Right Connector:** Choose the appropriate connector type and size for the wire gauge.

- **Crimp the Connector:** Use a crimping tool to securely attach the connector to the wire. Ensure a solid mechanical and electrical connection.

- **Inspect the Connection:** Check the crimp to ensure it is secure and free of loose strands.

6.2.4 Using Heat-Shrink Tubing

Heat-shrink tubing provides additional insulation and protection for wire connections:

- **Cut the Tubing:** Cut a piece of heat-shrink tubing slightly longer than the exposed connection.

- **Slide Over the Connection:** Slide the tubing over the connection before making the final crimp or solder.

- **Apply Heat:** Use a heat gun to evenly shrink the tubing around the connection. Ensure it adheres tightly to the wire and connector, creating a watertight seal.

6.2.5 Soldering Connections

Soldering provides a strong and durable connection, especially for critical circuits:

- **Prepare the Wire:** Strip the insulation and twist the wire strands together.

- **Apply Flux:** Apply a small amount of flux to the wire to promote good solder flow.

- **Heat the Connection:** Use a soldering iron to heat the wire and connector evenly.

- **Apply Solder:** Apply solder to the heated connection, allowing it to flow into the joint. Avoid applying solder directly to the iron.

- **Inspect the Joint:** Check for a smooth, shiny solder joint without excess solder or cold joints.

6.3 Mounting and Securing Components

Properly mounting and securing electrical components is crucial for ensuring their longevity and reliable operation. This section covers the methods for mounting different types of components and considerations for vibration and moisture.

Mounting Methods for Different Environments

6.3.1 Mounting Electrical Panels and Boxes

Electrical panels and boxes house fuses, breakers, and connections, requiring secure mounting:

- **Select Mounting Location:** Choose a location that is accessible for maintenance and away from potential water exposure.

- **Use Mounting Brackets:** Use mounting brackets or plates to secure the panels and boxes. Ensure they are aligned and level.

- **Fasten with Stainless Steel Hardware:** Use stainless steel screws or bolts to attach the panels and boxes securely to bulkheads or other structures.

6.3.2 Mounting Batteries

Batteries must be securely mounted to prevent movement and ensure safety:

- **Battery Trays and Boxes:** Use battery trays or boxes designed to fit the specific battery size.

- **Secure with Straps or Clamps:** Use straps or clamps to hold the battery firmly in place. Ensure they are tight enough to prevent movement but not so tight that they damage the battery casing.

- **Ventilation:** Ensure proper ventilation to prevent the buildup of hydrogen gas.

6.3.3 Mounting Electronic Devices

Electronic devices such as GPS units, radios, and displays require secure and accessible mounting:

- **Select Accessible Locations:** Mount devices where they can be easily operated and viewed.

- **Use Manufacturer's Mounting Kits:** Use the mounting kits provided by the manufacturer for a secure fit.

- **Route Cables Neatly:** Route power and data cables to the devices, securing them to prevent movement and chafing.

Vibration and Moisture Considerations

6.3.4 Protecting Against Vibration

Vibration can damage electrical components and connections over time:

- **Use Vibration Dampers:** Install vibration dampers or mounts for sensitive components such as electronics and control panels.

- **Secure Wiring:** Ensure all wiring is securely fastened and protected from vibration-induced chafing.

- **Inspect Regularly:** Regularly inspect connections and components for signs of wear or damage due to vibration.

6.3.5 Protecting Against Moisture

Moisture is a significant concern in marine environments, potentially causing corrosion and electrical failures:

- **Use Waterproof Connectors:** Use waterproof connectors and terminals, especially in areas prone to moisture exposure.

- **Seal Penetrations:** Seal any hull or deck penetrations with marine-grade sealant to prevent water ingress.

- **Install Drip Loops:** Create drip loops in wiring runs to prevent water from traveling along the wire into connectors or components.

- **Use Conformal Coating:** Apply conformal coating to sensitive electronic circuit boards to protect them from moisture and corrosion.

This chapter has provided a comprehensive guide to the installation and construction of sailboat electrical systems. By understanding the essential tools and materials, mastering proper wiring techniques, and learning the best practices for mounting and securing components, you can ensure a safe, reliable, and efficient electrical system on your sailboat. These detailed step-by-step explanations are designed to help you achieve professional-quality installations that enhance the performance and longevity of your electrical system, providing peace of mind and a better sailing experience.

CHAPTER SEVEN

System Integration and Testing

System integration and testing are critical steps in ensuring that all components of your sailboat's electrical system work seamlessly together. Proper integration ensures compatibility between systems, while thorough testing and troubleshooting help identify and resolve any issues before they become problematic. Regular maintenance is essential to keep your electrical system in optimal condition and extend its lifespan.

7.1 Integrating Multiple Systems

Integrating multiple systems involves ensuring that various electrical components and subsystems work harmoniously. This includes addressing compatibility issues and establishing reliable communication networks.

Ensuring Compatibility Between Systems

7.1.1 Assessing Compatibility

Before integrating different systems, it's crucial to assess their compatibility:

- **Voltage and Current Ratings:** Ensure that the voltage and current ratings of components match. For example, a 12V device should be used in a 12V system.

- **Communication Protocols:** Verify that the communication protocols used by different devices are compatible (e.g., NMEA 2000, Ethernet).

- **Physical Connections:** Check that connectors and cables are compatible. Use appropriate adapters if necessary.

7.1.2 Interfacing Different Systems

Different systems often need to interface with each other. Here's how to achieve seamless integration:

- **Use Interface Modules:** Interface modules or converters can bridge the gap between systems with different communication protocols.

- **Common Ground:** Ensure all systems share a common ground to prevent ground loops and electrical noise.

- **Synchronization:** Synchronize clocks and data between systems to ensure accurate and timely information exchange.

Communication Networks (NMEA 2000, Ethernet)

7.1.3 NMEA 2000 Network

NMEA 2000 is a robust marine communication network standard that allows multiple marine devices to communicate over a single data backbone.

- **Components:** The NMEA 2000 network consists of a backbone cable, T-connectors, drop cables, terminators, and devices.

- **Installation:**

    1. **Plan the Network Layout:** Design the network layout, ensuring that the backbone runs through accessible areas and devices are connected via drop cables.

2. **Install the Backbone:** Securely mount the backbone cable along the planned route, using cable clamps to prevent movement.

3. **Add T-Connectors:** Install T-connectors along the backbone where devices will be connected.

4. **Connect Drop Cables:** Run drop cables from the T-connectors to the devices, securing them with cable ties.

5. **Terminate the Backbone:** Install terminators at both ends of the backbone to prevent signal reflection and ensure proper communication.

6. **Power the Network:** Connect the network to a 12V power source using a power T-connector.

7. **Test the Network:** Power on the network and verify that all devices communicate properly.

7.1.4 Ethernet Network

An Ethernet network provides high-speed data communication for devices such as computers, cameras, and entertainment systems.

- **Components:** The Ethernet network consists of Ethernet cables, switches, routers, and devices.

- **Installation:**

    1. **Plan the Network Layout:** Design the network layout, ensuring that Ethernet cables run through accessible areas and devices are connected via network ports.

2. **Install Cables:** Securely mount Ethernet cables along the planned route, using cable clamps to prevent movement.

3. **Install Switches and Routers:** Mount switches and routers in accessible locations, ensuring proper ventilation.

4. **Connect Devices:** Run Ethernet cables from the switches and routers to the devices, securing them with cable ties.

5. **Configure the Network:** Configure the switches and routers for proper network operation, including IP addressing and security settings.

6. **Test the Network:** Power on the network and verify that all devices communicate properly.

7.2 Testing and Troubleshooting

Testing and troubleshooting are essential to ensure that the electrical system operates correctly and reliably. This section covers pre-installation testing, troubleshooting common problems, and systematic testing procedures.

Pre-Installation Testing of Components

7.2.1 Testing Electrical Components

Before installing electrical components, it's crucial to test them to ensure they function correctly:

- **Visual Inspection:** Inspect each component for any visible damage or defects.

- **Voltage Testing:** Use a multimeter to check the voltage rating of components to ensure they match the system requirements.

- **Continuity Testing:** Check the continuity of wires and connections to ensure there are no breaks or shorts.

- **Functional Testing:** Power up components individually to verify their operation. For example, test lights by connecting them to a power source.

7.2.2 Testing Communication Devices

For communication devices, ensure they can send and receive data correctly:

- **GPS Units:** Test GPS units to ensure they acquire satellite signals and display accurate position data.

- **Radios:** Test VHF and SSB radios to ensure they transmit and receive signals clearly.

- **AIS:** Test AIS transceivers to ensure they send and receive vessel information correctly.

Troubleshooting Common Problems

7.2.3 Identifying Issues

When troubleshooting, start by identifying the problem and narrowing down the possible causes:

- **No Power:** If a device does not power on, check the power supply, fuses, and connections.

- **Intermittent Operation:** For devices that work intermittently, check for loose connections, damaged wires, or corrosion.

- **Communication Failures:** If communication devices do not work, check the network connections, terminators, and settings.

7.2.4 Systematic Troubleshooting

Use a systematic approach to troubleshoot issues:

- **Check Power Supply:** Verify that the power supply is delivering the correct voltage and current.

- **Inspect Connections:** Check all connections for tightness, corrosion, and proper contact.

- **Test Individual Components:** Isolate and test individual components to identify faulty devices.

- **Check Network Integrity:** For communication issues, test the integrity of the network using diagnostic tools.

7.2.5 Using Diagnostic Tools

Diagnostic tools can help identify and resolve issues:

- **Multimeter:** Use a multimeter to measure voltage, current, and resistance.

- **Network Analyzers:** Use network analyzers to test the integrity of communication networks (NMEA 2000, Ethernet).

- **Oscilloscope:** An oscilloscope can help visualize electrical signals and diagnose complex issues.

7.3 Regular Maintenance

Regular maintenance is essential to keep the electrical system in optimal condition and prevent problems. This section covers maintenance schedules, checklists, and preventive measures.

Maintenance Schedules and Checklists

7.3.1 Developing a Maintenance Schedule

Create a maintenance schedule that outlines regular inspections and tasks:

- **Daily Checks:** Perform daily checks of essential systems, such as navigation lights, radios, and battery levels.

- **Weekly Checks:** Conduct weekly checks of critical connections, fuses, and circuit breakers.

- **Monthly Checks:** Perform monthly inspections of wiring, connectors, and terminals for signs of wear or corrosion.

- **Annual Checks:** Conduct annual maintenance of major components, such as batteries, charging systems, and communication devices.

7.3.2 Maintenance Checklists

Use maintenance checklists to ensure all tasks are completed systematically:

- **Daily Checklist:**
    - Check battery voltage levels.
    - Verify operation of navigation and communication systems.
    - Inspect critical connections for tightness.

- **Weekly Checklist:**
    - Test all lights and signals.
    - Inspect and clean battery terminals.
    - Check for signs of moisture or corrosion in electrical compartments.
- **Monthly Checklist:**
    - Inspect wiring for wear or damage.
    - Test the functionality of all circuit breakers and fuses.
    - Check the operation of the bilge pump and alarms.
- **Annual Checklist:**
    - Perform a load test on batteries.

- Inspect and clean solar panels and wind generators.
- Test the integrity of the communication network (NMEA 2000, Ethernet).

Preventive Measures to Extend System Life

7.3.3 Protecting Against Corrosion

Corrosion is a major issue in marine environments. Take these preventive measures:

- **Use Corrosion-Resistant Materials:** Use marine-grade, tinned copper wire, and stainless steel hardware.

- **Seal Connections:** Use heat-shrink tubing and waterproof connectors to seal connections against moisture.

- **Apply Corrosion Inhibitors:** Apply corrosion inhibitors to terminals and connectors.

7.3.4 Managing Vibration and Movement

Vibration and movement can cause wear and damage:

- **Secure Wiring:** Use cable ties, clamps, and conduit to secure wiring and prevent movement.

- **Dampers:** Install vibration dampers for sensitive equipment.

- **Regular Inspection:** Regularly inspect connections and wiring for signs of wear or chafing due to vibration.

7.3.5 Maintaining Batteries

Proper battery maintenance extends their life and ensures reliable operation:

- **Regular Charging:** Keep batteries fully charged to prevent sulfation.

- **Water Levels:** Check and maintain water levels in lead-acid batteries.

- **Clean Terminals:** Clean battery terminals regularly to prevent corrosion.

- **Load Testing:** Perform load tests annually to assess battery health.

7.3.6 Updating Firmware and Software

Keep firmware and software for electronic devices up to date:

- **Regular Updates:** Check for and install updates for GPS, chartplotters, radios, and other electronic devices.

- **Backup Configurations:** Backup configurations before updating to prevent data loss.

This chapter has provided a comprehensive guide to system integration, testing, and maintenance of your sailboat's electrical system. By ensuring compatibility between systems, establishing reliable communication networks, and conducting thorough testing and troubleshooting, you can achieve a seamless and dependable electrical setup. Regular maintenance, supported by detailed schedules and checklists, along with preventive measures against corrosion, vibration, and other potential issues, will help extend the life of your electrical system and ensure it operates efficiently and reliably for years to come.

CHAPTER EIGHT

Case Studies and Advanced Topics

In this final chapter, we will explore real-world examples of successful sailboat electrical system installations, delve into advanced systems and innovations shaping the future of sailboat technology, and provide additional resources to help deepen your understanding and aid in your projects. By learning from actual installations and staying updated on the latest advancements, you can ensure your sailboat's electrical system is both cutting-edge and reliable.

8.1 Real-World Examples

Case studies of successful installations offer invaluable insights into best practices, challenges faced, and lessons learned. These examples highlight different approaches to sailboat electrical system design and implementation.

Case Study 1: Coastal Cruiser Retrofit

8.1.1 Project Overview

- **Boat Type:** 36-foot coastal cruiser

- **Objective:** Retrofit the electrical system for extended coastal cruising

- **Key Components:** Lithium-ion battery bank, solar panels, wind generator, NMEA 2000 network

8.1.2 Planning and Design

- **Assessing Needs:** The owners planned to use the boat primarily for extended coastal cruising, requiring a robust electrical system to support navigation, communication, and comfort.

- **Budgeting:** A detailed budget was created, balancing cost and quality. High-priority items included a lithium-ion battery bank and renewable energy sources.

8.1.3 Installation Process

- **Batteries:** A 400Ah lithium-ion battery bank was installed, chosen for its high energy density and long lifespan. Custom battery boxes were built to secure the batteries.

- **Charging Systems:** Two 150W solar panels were mounted on the stern arch, and a wind generator was installed on a dedicated pole. An MPPT charge controller was used to optimize solar charging.

- **NMEA 2000 Network:** The existing navigation system was upgraded to NMEA 2000, connecting the GPS, radar, AIS, and depth sounder for seamless data sharing.

8.1.4 Challenges and Solutions

- **Space Constraints:** Limited space for additional batteries and components was a challenge. Custom mounts and careful planning allowed for efficient use of space.

- **System Integration:** Integrating the new NMEA 2000 network with legacy devices required interface modules and firmware updates.

8.1.5 Outcome and Lessons Learned

- **Outcome:** The retrofit was successful, providing reliable power for extended cruising. The owners reported significant improvements in energy efficiency and system reliability.

- **Lessons Learned:** Thorough planning and flexible problem-solving were crucial. The importance of high-quality components and proper integration was underscored.

Case Study 2: Offshore Racing Yacht

8.1.6 Project Overview

- **Boat Type:** 50-foot offshore racing yacht

- **Objective:** Upgrade the electrical system for performance and reliability in offshore conditions

- **Key Components:** High-capacity AGM battery bank, advanced navigation and communication systems, lightweight wiring

8.1.7 Planning and Design

- **Assessing Needs:** The racing yacht required a high-performance electrical system to support navigation, communication, and safety systems, with a focus on weight reduction.

- **Budgeting:** Given the competitive nature, a higher budget was allocated for top-tier components and professional installation.

8.1.8 Installation Process

- **Batteries:** A 600Ah AGM battery bank was installed, chosen for its reliability and performance under heavy loads. The batteries were strategically placed to optimize weight distribution.

- **Wiring:** Lightweight, high-performance marine-grade wire was used throughout to minimize weight without compromising durability.

- **Navigation and Communication:** State-of-the-art GPS, radar, AIS, and satellite communication systems were installed, all integrated via an NMEA 2000 network.

8.1.9 Challenges and Solutions

- **Weight Management:** Balancing the need for robust systems with weight constraints required careful selection of components and materials.

- **Harsh Conditions:** Ensuring system reliability in harsh offshore conditions led to the use of ruggedized components and extensive waterproofing measures.

8.1.10 Outcome and Lessons Learned

- **Outcome:** The upgraded system performed flawlessly during races, providing reliable power and seamless data integration.

- **Lessons Learned:** Investing in high-quality components and professional installation paid off in terms of performance and reliability.

Rigorous testing under realistic conditions was essential.

8.2 Advanced Systems and Innovations

The future of sailboat electrical systems is evolving with new technologies and innovations that enhance performance, convenience, and sustainability. This section explores some of these advancements.

Smart Systems and Automation

8.2.1 Introduction to Smart Systems

Smart systems use advanced technology to automate and optimize electrical system management:

- **Smart Batteries:** Intelligent battery management systems (BMS) monitor and manage battery health, optimizing charging and discharging cycles to extend battery life.

- **Integrated Monitoring:** Centralized monitoring systems display real-time data from all electrical components on a single interface, accessible via smartphones or tablets.

- **Automated Controls:** Smart controllers automatically manage lighting, climate control, and other systems based on pre-set conditions or user preferences.

8.2.2 Benefits of Automation

- **Efficiency:** Automated systems optimize energy use, reducing waste and improving overall efficiency.

- **Convenience:** Remote monitoring and control provide convenience and peace of mind, allowing users to manage their systems from anywhere.

- **Safety:** Smart systems can detect and respond to faults or unusual conditions, enhancing safety.

8.2.3 Implementing Smart Systems

- **Selection of Components:** Choose components that are compatible with smart technologies and can be integrated into a unified system.

- **System Design:** Plan the layout and integration of smart components, ensuring seamless communication and control.

- **Installation and Configuration:** Install smart components according to manufacturer instructions and configure the system to suit your preferences.

Future Trends in Sailboat Electrical Systems

8.2.4 Renewable Energy Integration

Renewable energy sources are becoming increasingly important in sailboat electrical systems:

- **Solar Power:** Advances in solar panel technology are making them more efficient and affordable. Flexible panels and high-efficiency cells are particularly suitable for marine applications.

- **Wind Power:** Modern wind generators are quieter, more efficient, and easier to install. They provide a valuable supplementary power source, especially during long passages.

- **Hybrid Systems:** Combining solar, wind, and traditional charging methods (e.g., alternators) creates robust hybrid systems that ensure continuous power availability.

8.2.5 Energy Storage Innovations

Advancements in energy storage are transforming sailboat electrical systems:

- **Lithium-Ion Batteries:** Continuing improvements in lithium-ion battery technology are increasing energy density, reducing weight, and enhancing safety.

- **Solid-State Batteries:** Emerging solid-state batteries promise even greater energy density and safety, potentially revolutionizing marine energy storage in the future.

8.2.6 Advanced Communication Systems

Communication technologies are evolving to provide better connectivity at sea:

- **Satellite Internet:** Advances in satellite technology are making high-speed internet more accessible and affordable for sailboats, enabling reliable communication even in remote areas.

- **Mesh Networks:** Mesh networking technology allows devices to communicate directly with each other, improving reliability and range without relying on a single central router.

8.3 Appendices and Resources

To support your ongoing learning and projects, this section provides a glossary of terms, a list of suppliers and manufacturers, and recommendations for further reading and reference materials.

Glossary of Terms

8.3.1 Key Terms and Definitions

- **Ampere (A):** The unit of electric current.

- **Battery Management System (BMS):** A system that manages the charging and discharging of a battery, ensuring safe operation.

- **Circuit Breaker:** An automatic switch that stops the flow of electric current in a circuit as a safety measure.

- **Direct Current (DC):** An electric current flowing in one direction only.

- **Inverter:** A device that converts DC power into AC power.

- **Marine-Grade:** Materials and components designed to withstand the harsh marine environment.

- **National Marine Electronics Association (NMEA):** An organization that sets standards for marine electronics.

List of Suppliers and Manufacturers

8.3.2 Recommended Suppliers

- **Batteries:**
    - **Lithium-Ion:** Battle Born Batteries, Victron Energy
    - **AGM:** Lifeline Batteries, Odyssey Batteries
- **Solar Panels:** Renogy, SunPower
- **Wind Generators:** Rutland, Air Breeze
- **Wiring and Cables:** Ancor Marine, Blue Sea Systems
- **Communication Systems:** Garmin, Raymarine

Further Reading and Reference Materials

8.3.3 Recommended Books and Articles

- **Books:**

    o "The 12-Volt Bible for Boats" by Miner K. Brotherton and Edwin R. Sherman

    o "Boatowner's Mechanical and Electrical Manual" by Nigel Calder

    o "Marine Electrical and Electronics Bible" by John C. Payne

- **Articles:**

    o "Advances in Marine Battery Technology" – Practical Sailor

- "Integrating Renewable Energy Systems on Boats" – Cruising World
- "Smart Boat Technology: The Future of Boating" – Marine Electronics Journal

8.3.4 Online Resources

Websites:

Marine Electronics Journal: www.marineelectronicsjournal.com

- **Practical Sailor:** www.practical-sailor.com
- **Cruising World:** www.cruisingworld.com

Summary

This chapter has explored the integration and testing of sailboat electrical systems through real-world case studies, advanced system innovations, and provided a wealth of resources for further learning. By examining successful installations, understanding the benefits and implementation of smart systems, and staying informed about future trends, you can ensure your sailboat's electrical system is both state-of-the-art and reliable. The glossary of terms, list of suppliers, and recommended reading materials offer additional support to enhance your knowledge and skills in designing, constructing, and outfitting modern sailboat electrical systems.

www.ingramcontent.com/pod-product-compliance
Lightning Source LLC
Chambersburg PA
CBHW071458220526
45472CB00003B/846